BATTLES FOR WARGAMERS

THE 2nd PUNIC WAR

by

Terence Wise

A BELLONA BOOK
Published by
MODEL & ALLIED PUBLICATIONS LTD.
13/15 BRIDGE STREET
HEMEL HEMPSTEAD HERTS

1972

Printed in Great Britain by
Technical Print Services Limited
Nottingham

BATTLES FOR WARGAMERS

THE 2nd PUNIC WAR

by

Terence Wise

A BELLONA BOOK

Published by

MODEL & ALLIED PUBLICATIONS LTD.

13/15 BRIDGE STREET

HEMEL HEMPSTEAD HERTS

1972

Printed in Great Britain by
Technical Print Services Limited
Nottingham

INTRODUCTION

Many books have been written about wars and specific battles but to date none have been written especially for the wargamer. Therefore the aim of this series of books is not to compete with more academic studies of military history but simply to provide the wargamer with concise, basic facts about a set of battles within one theatre or campaign, thus providing him – or her – with the raw material from which to reconstruct battles from history.

The first three books to be written in this series are **The 2nd Punic War, The American Civil War 1862** and **World War II : The Western Desert.** These cater for three of the most popular eras studied by the wargaming fraternity and it is hoped to expand the field in later books to cover the other major wars in which so many wargamers are interested.

As much as possible I have avoided the major battles which are already very well documented and have sought to include some of the smaller battles, which I personally consider more readily adapted to the limits of the wargames table. Information on these less famous but still important battles is also much harder to find and I hope by including such battles I will save players many frustrating hours of research and at the same time perhaps encourage them to widen their knowledge of the art of war. A list of my main sources is provided for those who wish to unearth more information than the scope of these books provides room for, or for those who wish to obtain a better understanding of the over-all scene.

All the books in this series follow a general plan : name, place and date; the aim; forces engaged, including all arms and services, forces in support and reinforcements if applicable; a description of the terrain, including man-made defences but avoiding duplication of details already shown by the maps; any major 'outside' events which influenced the decisions of the commanders; an account of the course of the actual battle, since the mistakes and results of these battles provide lessons and ideas for players; a summing up of the battle, pointing to moves which decided the outcome; and a list of casualties, including guns, tanks etc. The battles may be fought individually or as a set of six in a campaign: in some cases, notably the 20th century battles, it is often best to divide a battle into several games in order to develop the situation to its full extent.

Each account is accompanied by a map and a photograph of the battle as reconstructed for a wargame. Information about contours was especially hard to come by and, rather than have only partial marking of heights in fact, I have divided the high ground into four levels at the most, and these can be represented on the table by a corresponding number of layers of whatever material is being used to simulate high ground. The maps are complemented by the photographs which, usually taken in the early stages of the reconstructed battles, show the scene as if from a point of high ground and illustrate more clearly some of the possible alternative moves which could have been made.

Finally, brief but condensed guides have been supplied to organisation, equipment and uniforms, so as to provide a starting point for those who are venturing into a particular period for the first time.

I hope all wargamers will find fighting these battles as interesting as I myself found the studying and refighting of them.

Terence Wise

PAINTING GUIDE

Nationality	Tunic	Cloak	Helmet	Shield	Body Armour
Roman hastati	*	Dark red or brown	Iron with purple or black plume	Hide, iron or bronze boss & edge†	Bronze breast plate or iron chain mail
Roman principes	*	Dark red or brown	Iron	Hide, iron or bronze boss & edge †	Bronze breast plate or iron chain mail
Roman triarii	*	Dark red or brown	Iron	Hide, iron or bronze boss & edge †	Bronze breast plate or iron chain mail
Roman velites	Unbleached (yellowish white)	Dark red or brown	Iron	Hide, iron or bronze boss & edge †	None
Carthaginian cavalry	Purple	–	Bronze	2′ round, gay colours	Bronze corslet with inlay
infantry	Red	–	Bronze	3′ round, hide with metal boss & edge	None
Numidian cavalry	Unbleached	–	None	Small, round, of hide	None
infantry	Unbleached	–	None	Small, round, of hide	None
Libyans	Red leather	–	None	Round, hide	None
Spanish infantry light	White with red borders	–	None	2′ round, gay colours	None
infantry heavy	White with red borders	Usually Black	None	Gallic design	None
Gallic cavalry	Yellow, red, purple, browns tartans or stripes	Red or blues	Bronze	Gallic design decorated, the metal enamelled	None
infantry		Black, tartans or stripes	Bronze		None

* Choice of colour was left to the Legion's commander, but it should be borne in mind that during service in other countries only local dyes and materials would be available to replace worn clothing. Colours were usually dictated by the dyes known to Man at that time and would have been mostly the various greens, browns, blues, saffron yellow, red and purple.

† In some Legions the cohorts were distinguished by their shields, which were painted in different colours.

SOURCES

Cottrell, Leonard	Enemy of Rome	Evans Bros., 1960
Fuller, J.F.C.	Decisive Battles of the Western World	Eyre & Spottiswoode 1954
Livy, translated Aubrey de Selincourt	The War with Hannibal	Penguin 1965
Plutarch, translated Ian Scott-Kilvert	Makers of Rome	Penguin 1965
Scullard, H.H.	Scipio Africanus in the Second Punic War	Cambridge University Press 1930
Shuckburgh, E.S.	History of Rome	Macmillan 1916

CONTENTS

COLOUR KEY

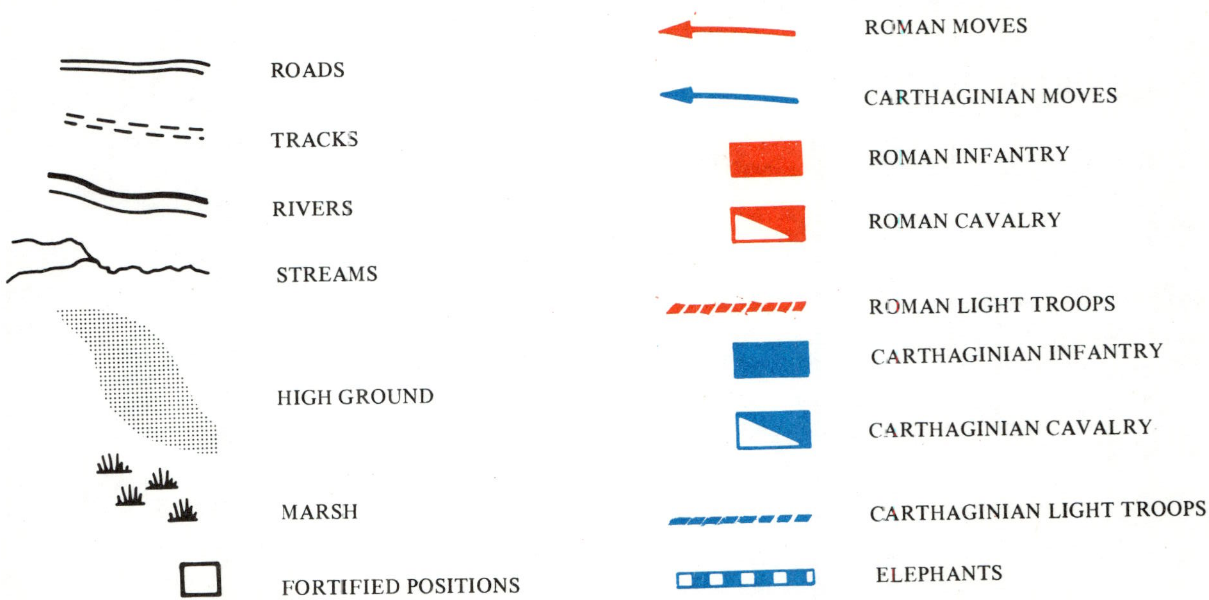
ROADS
TRACKS
RIVERS
STREAMS
HIGH GROUND
MARSH
FORTIFIED POSITIONS
ROMAN MOVES
CARTHAGINIAN MOVES
ROMAN INFANTRY
ROMAN CAVALRY
ROMAN LIGHT TROOPS
CARTHAGINIAN INFANTRY
CARTHAGINIAN CAVALRY
CARTHAGINIAN LIGHT TROOPS
ELEPHANTS

BATTLE OF THE METAURUS

Northern Italy, Spring 207 B.C.

Under the leadership of Fabius the Romans had slowly worn Hannibal down, reconquering Samnium and Syracruse in 212, and Capua, Brundisium and Tarentum between 211 and 209. The fall of Capua had cut Hannibal off from Campania, the recapture of Tarentum from Apulia, and by 209 he was confined in southern Lucania and Bruttium – the toe of Italy. To free him from this encirclement, in 209 Carthage ordered Hasdrubal to Italy.

ROMAN FORCE. Livius' army of two legions (10,000) and an equal number of Italian allies; Porcius' army of the same strength; 6,000 infantry and 1,000 cavalry of Nero's army. Total : 47,000. The number of cavalry with Livius and Porcius is not mentioned but perhaps 6,000 of the total force could have been cavalry.* Before the battle the three forces were united, then re-divided into centre and two wings.

CARTHAGINIAN FORCE. 14,000 veteran Spanish and African heavy infantry trained to fight in the Roman style; 8,000 Ligurians recruited in North Italy; 1,000 Numidian cavalry; 10 elephants; 10,000 Gauls, some of them cavalry, recruited in South West Gaul and from the Arverni and other tribes during the crossing of the Alps. Total : 33,000 with about 3,000 of that number cavalry.*

TERRAIN. Details of the exact area of the battle are rather limited and some of the contours are therefore missing from the map. We know that the Gauls were stationed on rough, hilly ground, protected on their left flank by the Metaurus itself, and in front by a steep ravine. The remainder of the battle front was confined, the Carthaginian army being drawn up 'along a narrow front' so that the centre and right wing were more than usually deep in formation.

After an abortive attempt to take Placentia, Hasdrubal advanced on Ariminium (Rimini), driving the armies of Porcius and Livius before him, and sent a message to Hannibal that he intended to march on Ariminium then join his brother in Umbria, on the Adriatic side of Italy. Unfortunately his messengers were captured and armed with this information, Nero, guarding Hannibal in the south, was able to slip away with 7,000 men to join Livius, leaving the bulk of his army watching Hannibal, who was to wait in vain for news of Hasdrubal.

Nero covered the 250 miles in seven days and slipped into the camp of Livius, his men sharing the tents of the other legions. The next day the Romans drew up for battle but Hasdrubal, noting the increase in numbers, refused to be drawn and sent out patrols to try to discover who the reinforcements were. The patrols were unable to find any enlargement of the camp, but reported hearing duplicated trumpet calls in the camp of Livius. With his long experience of fighting the Roman army, Hasdrubal guessed Nero had arrived – what had happened to Hannibal and his army? How could Nero march north, unless Hannibal had been crushed? Fearing the worse, Hasdrubal withdrew his troops into their camps and that night moved away to cross the Metaurus and put a safe distance between himself and the Romans, whose camp had been less than half a mile from his own.

In the confusion of breaking camp and moving off in darkness, Hasdrubal's guides managed to escape and, losing their way, the Carthaginians followed the tortuous river for most of the

**Cavalry do not appear to have played much of a part in this battle, either because of the lack of room to manoeuvre or because of their small numbers. Their numbers and role are therefore my own interpretation and are not confirmed by any of the sources. Light troops are also ignored by the historians, but here there are no clues at all as to their part in the battle.*

night, putting little distance between themselves and the Romans. At last Hasdrubal called a halt, planning to wait for day break and then find a ford, but in the morning he was unable to find a suitable crossing place and the further up-river he went the higher the banks became until most of the morning had been wasted and the Romans were able to catch up with him.

Nero arrived first with the cavalry, then the light troops came on the scene and these forces began to harass the weary Carthaginian army. Hasdrubal abandoned his attempt to escape and began to fortify a camp on the high ground above the river, but before he could make much progress Livius arrived with the legions – not in column of route, but formed up ready for battle. Hasdrubal had no alternative but to turn and fight. During the night his Gallic contingent had begun to disintegrate, some deserting under cover of darkness, others becoming drunk, more collapsing from exhaustion. Those who were fit for battle Hasdrubal posted on his left flank, in a strong natural position, where they were almost unassailable.

The battle began in the centre with a charge by the elephants, which caused confusion in the Roman ranks and forced their line back for a while. On the Carthaginian right battle was also joined, but on the other wing, despite several attempts at an assault by Nero, the Gauls remained safe on their hill.

At first the advantage was with the Carthaginians in the main battle, but as the noise of battle became greater some of the elephants got out of control and began to inflict as many casualties on their own side as the enemy. The mahouts were equipped with a chisel and mallet and by driving the chisel between head and neck some of the elephants were killed. Then the battle raged on, the Romans regaining the ground they had lost, and advantage going to neither side for a long time.

It must be assumed that, with their superiority in cavalry, the Romans were able to gain mastery of the open flank – their left. Meanwhile Nero had become impatient at his enforced idleness on the other wing and, leaving part of his force to face the Gauls, he slipped away with the remainder down behind Porcius and Livius, so swiftly that most of the Romans were not aware of his manoeuvre. The sources call this force 'several cohorts' but to achieve what he did, Nero must have taken at least a third of his force, one legion. Suddenly this force appeared on the flank of the heavily engaged Spanish and African infantry and, lapping round their line, attacked them from flank and rear, driving them into the Ligurians and pushing on to begin killing some of the Gauls on the far left.

It was now midday and the Carthaginians were beginning to feel the effects of their night march. Inspired by Hasdrubal's leadership they fought for a little longer, but, losing heavily, the army began to break up and when it became obvious the day was lost Hasdrubal galloped into a Roman cohort and died fighting.

CONCLUSIONS. The battle ended Hannibal's last chance to break Rome's power in Italy. Recognising this, Hannibal collected as many troops as possible and withdrew into Bruttium, where he remained for three years before being recalled to Carthage. From now on the war would be fought in Spain and Africa.

CASUALTIES. Roman : 8,000 killed. **Carthaginians** : 10,000 killed (mostly Spanish and African heavy infantry), and 2,000 captured. 6 elephants were killed, the rest captured. Most of the Gauls and Ligurians, and those Spaniards who survived the battle, managed to flee across the Metaurus and were allowed to escape so as to carry news of the defeat to their homes and discourage further allegiance to Carthage.

BATTLE OF THE METAURUS

BATTLE OF LAKE TRASIMENE

Central Italy, April 217 B.C.

Having broken into Italy and defeated the consul Sempronius at Trebbia during the winter of 218-17, in the spring Hannibal deliberately placed himself between the armies of the new consuls, Servilius and Flaminius, with the intention of dealing with them separately.

ROMAN FORCE. Most sources, and the casualty figures, give a total of 40,000 under Flaminius but the new armies were composed of 10,000 survivors from Trebbia, four newly raised legions (20,000), and presumably an equal number of Italian allies – the normal practice when raising an army. This gives a total of only 60,000, which was then divided equally between the two consuls. I prefer the 30,000 arrived at by this calculation to the usual 40,000. Cavalry are not mentioned with Flaminius, but would probably not have exceeded 4,000.

CARTHAGINIAN FORCE. 20,000 Spanish and African infantry including light troops and Balearic islanders; 8,000 Cisalpine Gauls; 6,000 Numidian cavalry under Maharbal; 4,000 Gallic cavalry. Total: 38,000 under Hannibal. (All the elephants except one had been lost either at Trebbia or in the abortive attempt to cross the Apennines during the winter to recruit the aid of the Etruscans.)

TERRAIN. East of the village of Passignano the mountains of Cortona slope down to Lake Trasimene, leaving only a narrow corridor for the road between mountains and lake. By the village a spur of the hills projects right down to the lake, while more spurs to the East again reduce the width of the corridor. At this eastern point the road climbs up a narrow defile to give a view of a wide plain. Beyond the plain can be seen more hills.

The armies of the consuls had been stationed near the line along which Hannibal was expected to advance: Servilius on the Adriatic coast, Flaminius at Arretium (Arezzo) on the western flank of the Apennines overlooking the Tuscan plain. Learning that Flaminius was at Arretium and that he had the reputation of being arrogant, Hannibal led his army past that city and began pillaging and burning the countryside between Cortona and Lake Trasimene, with the object of drawing Flaminius out to protect the land of Rome's allies. Flaminius' staff urged him to send cavalry and light troops to stop these depredations, but to wait for Servilius before attempting to engage Hannibal in a pitched battle. However, the sight of smoke rising from farms and villas was too much for Flaminius and he ordered the army out in pursuit, following Hannibal eastwards along the road which skirted the northern shore of Lake Trasimene.

The Carthaginian army camped at the eastern end of the corridor described above, with the Romans about four miles in their rear. Flaminius had conducted a forced march to achieve this narrowing of the gap between the two forces, and the Romans made camp so late in the evening that there was no opportunity to make a reconnaissance of the ground ahead. But Flaminius was thinking only of overtaking Hannibal – he was sure that the next morning he would be able to catch the plunder-laden Carthaginians in column of march. At first light, therefore, the Roman army set out in pursuit of the Carthaginians, Flaminius welcoming the thick mist which rose from the lake on his right, for it would conceal his approach. But the hunter was just about to become the hunted.

Hannibal had broken camp even earlier than the Romans and placed his troops for the forthcoming battle. A 'rearguard' of the veteran Spanish and African infantry was posted on the road near the crest of the hill beyond which the open plain began, the light troops were sent to take up positions in the hills North of the lake, with the Gauls further West of them, and all the cavalry at the extreme end of this line so that, as soon as the Romans had advanced into the corridor, they could block the exit at Passignano.

Greek and Roman historians make special mention of the heavy mist which rose from the lake at sunrise. Hannibal had something like 30,000 troops 'concealed' on the hill sides above the lake. Some of these hills were wooded, but many of the troops were exposed on the hill sides, for we know they were able to see each other and were able to co-ordinate their attack. It is more likely that Hannibal knew of these mists and, taking a slight gamble, relied on them to conceal the bulk of his troops.

There was no sign of the enemy as the Romans entered the corridor and the legions pressed on eagerly, the thick mist enclosing them so that they could see nothing on their flanks and very little ahead. As the head of the long column began to climb the road out of the corridor they cleared the low-lying mist and saw what appeared to be the rearguard of the Carthaginian army ahead of them. Here the land began to open out and as the Roman advance guard of 6,000 men came on the enemy, who had halted to fight, they deployed and a hand to hand battle began. Behind them the column wound its way along the narrow, mist-shrouded road, hearing the clash of arms and cheers ahead. The enemy had been caught.

Suddenly the mist was pierced by the sound of trumpets, followed by the war cries of 30,000 men and the earth-shaking rumble of 10,000 charging horsemen. Strung out in a thin vulnerable line, with the lake as a barrier to their right, the Roman legionaries turned to face this unknown factor. Stones, arrows and javelins flew out of the mist to strike them down before they could protect themselves. Abruptly 20,000 screaming Gauls, Spaniards and Africans broke through the mist and were on them before they could form a line, before many could even draw their swords. They had barely time to begin to fight back when another shower of javelins smote their flank and rear and the terrible Numidians rode into their midst.

Hemmed in on all sides, their organisation completely shattered from the start, still the Romans fought bravely. There were no cohorts or even centuries, but each man fought where he stood, Flaminius and a band of his best men rushing from spot to spot, striving futilely to organise a line. For nearly two hours this battle raged, so fiercely that historians record that none of the combatants were aware of an earthquake which destroyed large areas of many Italian cities. Then Flaminius was killed by a Gaul, and with him died the flower of the army. Panic began and spread, and men turned wildly to escape. Finding no way out, many in desperation plunged into the lake in the forlorn hope of swimming across it to safety. Exhausted and weighed down by their equipment most of them drowned: those who survived struggled back towards the shore, only to be cut down by the Numidians, who rode their horses into the water to meet them.

Meanwhile the 6,000 men of the advance guard had been drawn further and further away in their fight with the Carthaginian 'rearguard' which finally broke and fled to the safety of the hills. The 6,000 suddenly found themselves alone, with no sign of either the Carthaginian or their own army. Behind them they could hear the shouts of struggling men and the din of battle. They reformed and moved to a crest from where they could look into the valley. At first the mist prevented them from seeing anything, but it was now three hours since dawn and as the heat of the sun increased so the mist dispersed and they saw scattered along the narrow corridor the bodies of some 15,000 of their comrades, the Carthaginian infantry stripping the corpses and the cavalry despatching the last survivors. The 6,000 marched quickly eastwards, hoping to escape, but the next day they were surrounded by the Numidian cavalry and forced to surrender.

CONCLUSIONS. For many years Rome had been undefeated. Trebbia had been reported to the people as a disputed and uncertain victory for Hannibal, but here there was no room for conjecture. The situation was considered so desperate that a dictatorship was set up, Fabius Maximus being elected. The road to Rome lay before Hannibal, but he didn't take it. His reasons can only be guessed. He had no seige train; his men and horses had suffered severely in the past six months and an outbreak of scurvy was a serious threat to his strength; a large proportion of his army were Gauls, who could only be held together by the prospect of loot and whose temperament would not stand the boredom of a long siege. Hannibal also realised the only way to break Rome was to rob her of her allies, and to further this end he released all allies taken in the battle, telling them he came to liberate them from Roman rule.

CASUALTIES. Roman : 15,000 killed, 15,000 captured, including the advance guard, and 10,000 scattered throughout Etruria who eventually made their way to Rome. This figure of 40,000 may be brought about by the large number of prisoners: the advance guard were Romans and were put in chains; the remaining 9,000 may well have been allies, released by Hannibal, and so confused with the 10,000 stragglers – surely too large a number to have escaped such a trap? **Carthaginians** : 2,500, of which 1,500 were Gauls, which suggests the fiercest fighting was towards the rear of the Roman column.

BATTLE OF LAKE TRASIMENE

BATTLE OF CANNAE

Southern Italy, 3rd August, 216 B.C.

During his dictatorship Fabius instituted a system of delaying tactics designed to wear down Hannibal's strength. When consuls were re-elected in 217 they continued this method of dealing with the invaders, but by 216, when Varro and Paullus were elected, public opinion was demanding a decisive battle. A new army of four double legions together with an equal number of Italian allies was therefore raised, the largest army Rome had ever put into the field, and the two consuls set out to seek a victory. Several times Hannibal tried to lure them into traps but the experience of Paullus defeated his plans. However, the inexperienced, impetuous and self-confident Varro – and the men themselves – continuously pressed for action. Finally Hannibal moved to Cannae, a Roman depot town, where he captured the supplies he badly needed, and deprived the Romans of the same. The Romans followed and made camp some six miles from the Carthaginians. Under the system of joint command employed when both consuls were with an army, the next day was Varro's turn to command.

ROMAN FORCE. Four double legions (40,000) and an equal number of Italian allies commanded by the previous consul Servilius: 2,000 Roman cavalry under Paullus: 4,000 allied cavalry under Varro. Total : 80,000 infantry, 6,000 cavalry.

CARTHAGINIAN FORCE. 20,000 Gauls and perhaps 5,000 Spaniards, including the light troops (javelin throwers and Balearic slingers) under Hannibal and his brother Mago: about 15,000 African heavy infantry, now armed like the legions with captured equipment, also commanded by Hannibal: 6,000 Numidian cavalry under Maharbal: 4,000 Gallic cavalry under Hasdrubal Gisgo. Total : 40,000 infantry, 10,000 cavalry.

TERRAIN. Both main camps were on the North bank of the river. The battle being fought on the South bank, the river must have been easily forded. Also, the Romans could not have been concerned about the slope they faced: the contours rose from 35′ at their initial position, to 70′, then 100′ where the Carthaginians waited. Another factor was a scorching wind, which swept across the bare, sandy plain and, whipping up clouds of dust, drove them into the faces of the advancing Romans.

Despite advice from Paullus, who didn't like the open ground coupled with the Carthaginians' superiority in cavalry, the next morning – August 1st – Varro ordered a general advance and moved the army toward the river, intending to deploy for battle. Hannibal sent forward his light-armed troops and cavalry, caught the Romans in line of march, and inflicted some casualties. Varro brought forward his cavalry and velites, then advanced some of the legionaries and eventually the Carthaginians were beaten back and the Romans took up position by the river. By now the day was nearing its end and camp was made at the main site indicated on the map.

The next day Paullus was in command and he secured the position as best he could, considering it inadvisable to either withdraw or to force a battle. He fortified the main camp and set up a smaller camp on the South side of the river, in which he installed a third of the army, using it to protect his foragers and attack those of the enemy. Hannibal had also formed a camp on the North bank and drew up his army for battle, but Paullus was not to be drawn and the day passed without incident except for harassing by the Numidians, who rode right up to the lesser Roman camp and infuriated the inactive legionaries.

In the morning Varro was back in command and, without consulting Paullus, he led out the army, joined with the legions on the South side of the river, and drew up in line of battle with the Roman legions nearest the river, those of the allies on the coastal flank. The legions were in their customary three lines but with the maniples much closer together than usual. The Carthaginian army took up position facing them, their line about the same length, but as the two armies advanced towards each other Hannibal allowed the Gallic and Spanish infantry in the centre to push forward so that, by graduating the position of the units on each side of the centre, a crescent was formed, with the convex towards the Romans.

The battle began with the exchange of missiles from the light troops and this continued indecisively for some time until some of the Gallic cavalry on the Carthaginian left joined in and routed the Roman velites here, pursuing them towards the river and cutting them down without mercy. This was followed by a cavalry mêlée on the same wing. Hemmed in by the infantry on one side and the river on the other, there was no room for manoeuvre and the result was a head-on collision with riders dragging each other from the saddle and fighting more or less as infantry. For a short time there was no result, then the outnumbered Romans were forced to yield and, with heavy losses, were driven along the river bank by the Carthaginians.

The Roman right had also become engaged in a mêlée with the Numidian cavalry who wheeled and darted in and out from all directions. One body of 500, with swords concealed beneath their tunics, threw down their arms and surrendered, only to turn on the rear of their captors later, causing heavy casualties and disorder in the rear. Meanwhile the general mêlée continued on this flank for a long time without a decision.

About the time the Roman cavalry broke, the legions met the advanced line of Spanish and Gallic infantry. The lighter armed Carthaginian centre soon gave way and gradually the legions pushed forward until the enemy's line had changed from convex to concave and many of the Spaniards and Gauls had begun to break and run. But intent on their own scene, following the standards and keeping a tight shield line, the legionaries had failed to realise that on each flank lay the African heavy infantry – unengaged. Shouting their war cries, these two divisions now swung to smash with a terrific shock into the flanks of the legions. The Africans were fresh, the Romans tiring and surprised. Caught on the flank, losing their formation in turning to meet this new threat, the Romans began to give ground.

But for Hannibal it was a finely balanced manoeuvre. On his right the cavalry fight was still undecided and his centre was beginning to break up. Then, just in time, Hasdrubal returned with the Gallic horsemen. At first he turned to help the Numidian cavalry, but the Roman allied cavalry, seeing this new body of troops approaching in their rear, at last gave up the fight and fled from the field, hotly pursued by the Numidians. Hasdrubal checked the advance of his Gauls, turned, and led them straight into the rear of the Roman legions. Seeing the tide turning, the Spanish and Gallic infantry in the Carthaginian centre now rallied and returned to the attack with new strength.

All was now lost for the Romans but for a time they fought stubbornly on, clamped in a tight circle of enemies which jammed them tighter and tighter together round their standards until they were hardly able to wield their weapons. Still they would not surrender. Paullus, who had returned with a handful of men to help lead the infantry, was killed, then Servilius. Eighty senators and 29 military tribunes were also killed, Varro had fled. Finally, with almost two thirds of their number killed by repeated charges from all sides, the legions broke and the once great army was reduced to a mass of leaderless fugitives, many of whom were cut down, although thousands did succeed in escaping northwards up the road or across country and were later rallied to the defence of Rome.

CONCLUSIONS. Although the way was wide open, Hannibal still did not advance on Rome, presumably because of the reasons listed under Lake Trasimene. In three great battles he had killed a fifth of Rome's manpower and reduced her to a state of collapse. yet the war was to drag on for another 14 years. Nevertheless this victory had achieved Hannibal's prime objective: Samnium, Lucania, Apulia, Campania and Capua all defected to the Carthaginians after Cannae and Hannibal soon controlled almost the whole of Southern Italy. Capua, second city after Rome, was especially important, heading a coalition of states now hostile to Rome and alone capable of supplying 30,000 infantry and 4,000 cavalry. Hannibal now had a source of men and supplies. On the debit side he had – for the first time – to divide his army to garrison these cities and protect his new allies, and Mago was therefore sent to Carthage to recruit even more men.

CASUALTIES. Roman : 50,000 killed, including more than half the cavalry; 4,000 captured. 2,000 more were captured by the Carthaginian cavalry the next day. The 10,000 in the two camps were offered surrender but refused. About 1,000 fought their way out, some 2,000 of the others were killed in the subsequent fighting, and the remainder then surrendered. **Carthaginians :** 4,000 Gauls, 1,500 Spanish and African infantry, 500 cavalry killed. A total of 6,000.

BATTLE OF CANNAE

BATTLE OF CANNAE

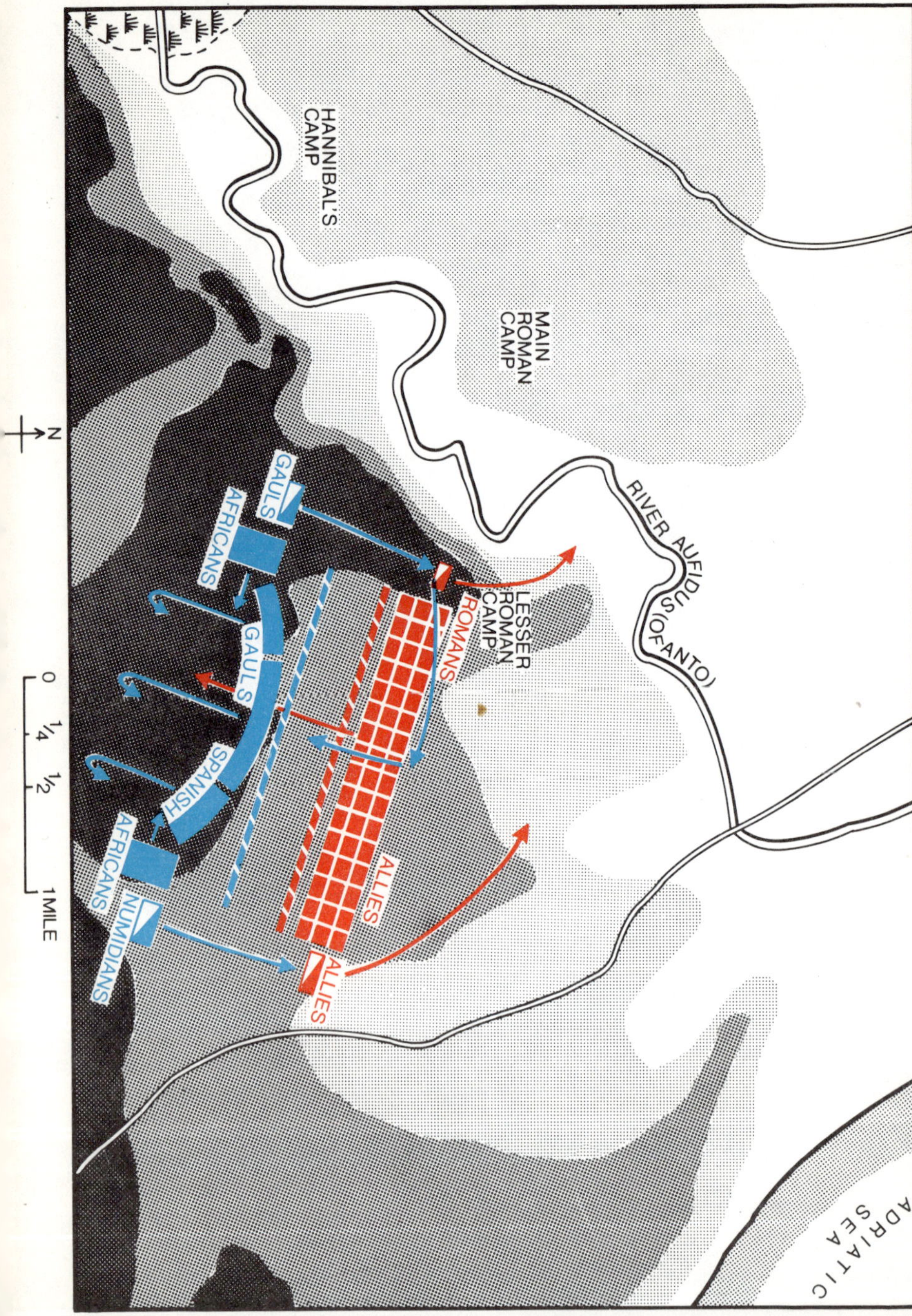

CANNAE N **1 man = 200**

A flat board was used here as the terrain played no part in the battle. The battle is seen at the time when the Carthaginian centre (nearest camera) is beginning to fall back and the two African 'divisions' are moving up to outflank the vastly superior numbers of Romans. The picture enables you to see the overwhelming numbers of the Romans – six ranks against three – which, when cramped by being forced into a horseshoe shape, acted against them. Note also the Carthaginian cavalry, victorious on the left and soon to be so on the right. Then all these men – representing 10,000 cavalry – smashed into the rear of the Legions and it was all over.

BATTLE OF BAECULA

Southern Spain, Spring 208 B.C.

Carthage had three powerful armies (under Hasdrubal Gisgo, Hasdrubal and Mago Barca) in Spain holding the country to the Carthaginian cause, but in 209 B.C. Scipio captured Carthage Nova (Cartagena) in South East Spain and during the following winter three of Carthage's strongest Spanish allies defected to the Romans. This was a serious blow to Hasdrubal Barca, engaged in raising money and men for his brother Hannibal, for he found himself cut off from his base, communications with the other two armies difficult, and his Spanish allies going over to the enemy. When Scipio confronted him at Baecula, therefore, although inferior in numbers, Hasdrubal was glad of a chance to fight – he desperately needed a decisive battle. Should he lose he could still reach Hannibal with some men and the money, recruiting more men on the way to Italy and leaving Scipio to Mago and the other Hasdrubal, but if he won the defection of the Spanish tribes would be reversed, Spain would be safe, and he could take a large army into Italy.

ROMAN FORCE. Roman citizens and Italian allies totalling about 21,000 (4 legions); some 4,000 seamen taken from the fleet and recruited into the legions; between 7,000 and 12,000 of the new Spanish allies under their Princes Indibilis and Mandonius; 3,000 cavalry (not mentioned in the battle but presumably opposing the Numidians). Total : between 35,000 and 40,000 depending on the strength of the Spanish allies, all commanded by Scipio.

CARTHAGINIAN FORCE. Reduced to about 25,000 by the loss of the Spanish allies, Hasdrubal's army consisted of about 15,000 Spanish and African heavy infantry; 7,000 Spanish and African light infantry and Balearic slingers; 3,000 Numidian cavalry. (The cavalry fought with the light troops for this battle and may well have fought on foot.) There were also about ten elephants.

TERRAIN. There is some doubt as to whether Hasdrubal's camp was on the height between the Arroyos de la Muela and Cañada Baeza, or the Cañada Baeza and del Matadero. The former is the larger and the terrace or ridge half way up the slope more suited to the description of the fighting. The height of the terrace here is 1,000′, that of the highest point about 1,350′, the latter also having a good size plateau. Hasdrubal set up his camp on the highest point, protected on each flank by the arroyos and in front by the sharply sloping rough ground. About half way down, where the slope levelled out into the wide terrace, he stationed his light troops, again protected on the flanks by the arroyos and in front by a steep bank, falling away to the river. The whole position was protected on the North and South by mountainous regions so that Scipio was forced to approach from the East, an approach commanded by Hasdrubal's position.

Hasdrubal had been collecting silver from the mines near Baecula (Bailen) but on the approach of Scipio's army he took up the naturally strong position described above. He needed a battle, but only on his own terms, and he took up this position in the hope that Scipio would throw his army against it and so cancel out his numerical superiority. If Scipio refused battle then Hasdrubal could wait until Mago and Hasdrubal Gisgo arrived on the scene.

Scipio made camp in the Arroyo de Tobarias and for two days each commander waited for a move in the other camp, but time was against Scipio. In addition to the possibility of the other Carthaginian armies arriving on the scene, he had been forced to make camp in the only strong position available, a spot where there was a shortage of water. On the third day, therefore, he decided to launch an assault on the strongpoint.

Firstly he sent one cohort to guard the road which led to Baecula, over the hill and towards

Linares, and another cohort up the Arroyo de Cañada Baeza on Hasdrubal's right. With the flanks thus guarded, he began to lead out his army and in the meantime sent the velites and a picked force of infantry to attack the Carthaginian light troops on the terrace. This force waded across the waist-deep Guadiel river and began to climb the slope without opposition, but as they came within range the Carthaginians assailed them with a hail of missiles. Loose stones were scattered about in plenty and these were also used in the exchange of missiles which continued for some time, the Romans hanging on grimly to what ground they could gain.

Meanwhile Scipio had brought his main army up to the river in close support of the advance guard. To Hasdrubal on the heights the Roman line of battle must have looked very strange, for Scipio had broken the rigid, mechanical tactics of the Roman army for the first time. The traditional Roman formation of hastati, principes and triarii relied on weight, and while it could advance or retire in an impenetrable line, it was almost impossible for it to change front, and any attempt to do so in the war so far had usually resulted in it being outflanked and severely mauled by the far more mobile Carthaginian forces. Baecula was important in that it marked a turning point in Roman tactics. During the winter Scipio had drilled his legions harshly and knew he could rely on them to manoeuvre in separate divisions instead of the familiar three lines. Now he drew up the legions in two bodies on the flanks, with 'some infantry and the light troops in the centre', then he sent forward 'the whole of his light troops to support the first attack and keep the enemy's attention directed to their front'.

No mention has been made of the Spanish allies in accounts of the battle but it is stated that Scipio used his best troops on the flanks. From this, and the quotes above, I would think that the light troops in his centre, which he now sent forward, were the Spanish infantry, stiffened by some cohorts of Romans. Under the impetus of these reinforcements the Roman line at last fought its way up to the level ground, where they soon began to cut up the Carthaginian light troops, who were not trained for hand to hand fighting, nor had the weapons for it.

Hasdrubal had until now kept his main force in the camp, hoping the ground would discourage a serious assault on his position, but seeing Scipio's main force entering the battle he realised the Roman commander was determined to storm the camp, regardless of loss. He now began to move forward his heavy infantry to help the weakening light troops below.

But Scipio had made another revolutionary move – the splitting of his army in the face of the enemy. Sending half the legions off to the left, up the Arroyo de Cañada Baeza, under Laelius, he led the remainder up the Arroyo de la Muela. The Carthaginian infantry began to move forward to their new position on the terrace, where they believed the main battle to be taking place, but Hasdrubal had despatched them too late. Before his heavy infantry could take up this position Scipio's legions arrived on their flank and threw them into disorder. Attempting to form a front to face this threat, the Carthaginians were next assailed on their other flank by Laelius. Their formation broken, the heavy infantry began to fall back to prevent encirclement.

For Hasdrubal there was no point in staying to fight to the bitter end. He had already sent his elephants and the gold and silver off towards the North, now he pulled out his heavy infantry and retreated, abandoning the remainder of his army. Driven back to the higher plateau and surrounded, these troops now began to surrender to the victorious Romans.

CONCLUSIONS. Hasdrubal had sacrificed his light troops and much of his cavalry in order to save his veteran heavy infantry and Scipio, who had not succeeded in surrounding the Carthaginian army and could not prevent the main body from withdrawing from the battle, achieved a tactical victory but suffered a strategic defeat. Hasdrubal had lost two fifths of his army but the remainder, the elephants and the vast sums of money he had collected to pay for mercenaries, were saved. He withdrew towards the Tagus, followed by only a small body of men who were to report on his movements. Scipio's orders were to hold Spain for Rome and he dared not follow Hasdrubal, leaving Mago and the other Hasdrubal in his rear. Hasdrubal was therefore able to cross into France and eventually make his way to Italy, unopposed, and recruiting a new army on the way.

CASUALTIES. Roman : Unknown. **Carthaginian :** 8,000 infantry, mostly Balearic, Spanish and African light troops, and nearly 2,000 cavalry killed and captured. The Spanish troops were released to encourage further defection to the Roman cause: the Africans were sold into slavery.

BATTLE OF BAECULA

BATTLE OF BAECULA

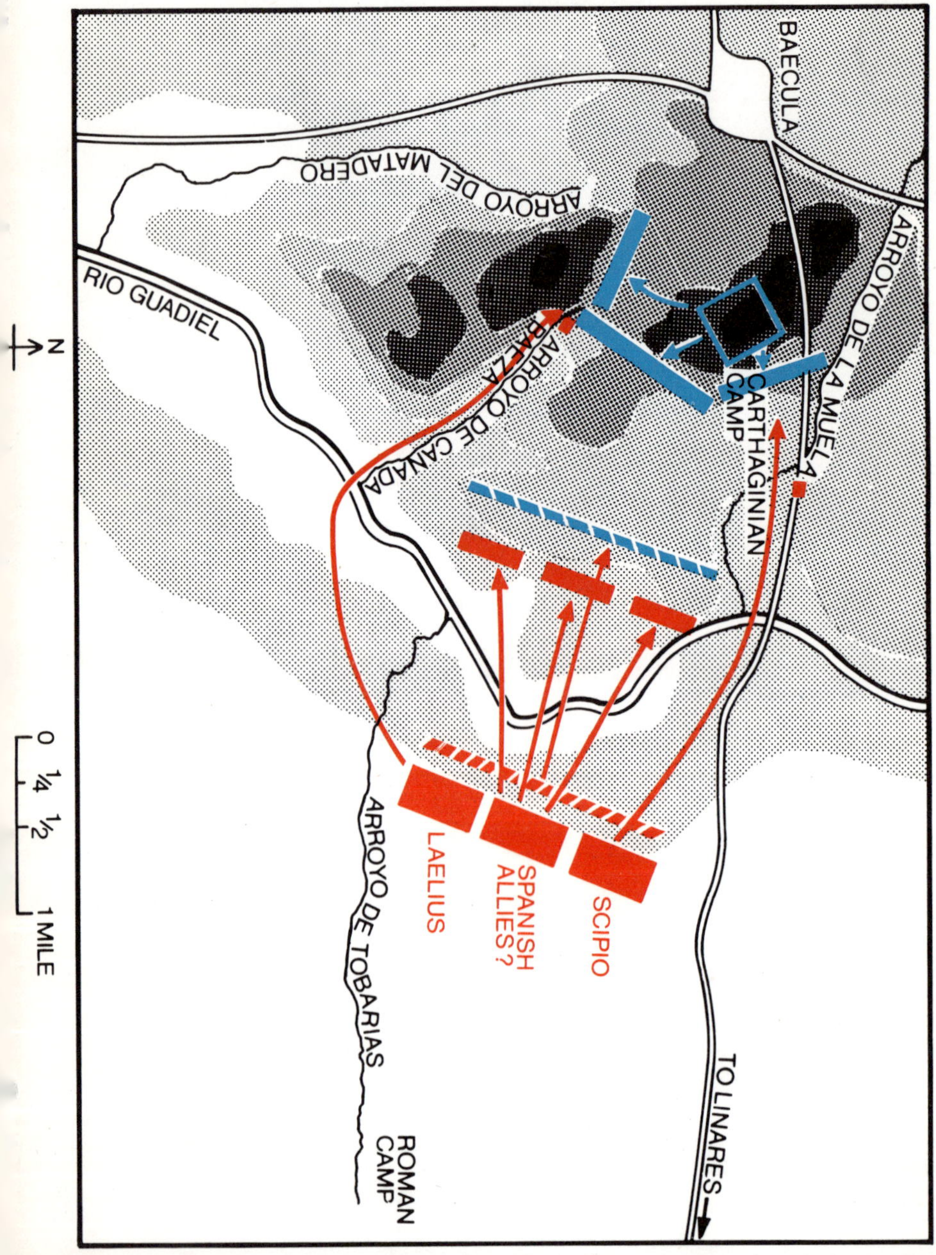

BAECULA — N — 1 man = 100, 1 elephant = 5

As the roads and arroyos themselves did not play a part in the action they have been left off the board. The scene shows only the main hill, with valleys on each side, and the river which the Romans had to cross. The two Roman columns are just about to nip into the Carthaginian flanks in a classic pincer movement – but too soon, for most of the heavy infantry is still too far back to be caught. In the foreground the Romans' Spanish allies have gained the terrace and are engaged in hand to hand fighting with the Carthaginian light infantry.

BATTLE OF ILIPA

Southern Spain, April 206 B.C.

When news of the battle of the Metaurus reached Hasdrubal Gisgo at Gades (Cadiz) he knew there was no longer any point in employing delaying tactics in Spain – Hannibal was still isolated in the toe of Italy and needed help. Hasdrubal decided to gamble everything on one decisive battle. If he won Spain and probably most of Italy would again come under Carthaginian control; if he was defeated then both Spain and Italy were lost.

ROMAN FORCE. 20,000 Romans and Italian allies; about 22,000 Spanish allies; 2,500 Roman and Italian cavalry; 500 Spanish cavalry. Total : 45,000.

CARTHAGINIAN FORCE. An army of 30,000 based on 10,000 African and Carthaginian infantry sent to Spain in 207, the remaining 20,000 being a mixture of the old Carthaginian army in Spain – Carthaginians, Libyans and Numidians – mercenaries from the Balearic Islands, and veteran Spanish allies. To this were added 20,000 new Spanish allies from Baetica. Total infantry : 50,000, Cavalry : 3,000 Numidians under Masinissa, 1,500 Carthaginians under Mago Barca. Total cavalry : 4,500 and 32 elephants. Total overall : 54,500.

TERRAIN. The battle took place on level ground between the two camps of the opposing forces, closest to the Carthaginian camp. The camps themselves were on slight slopes, nowhere exceeding 300′ in height.

Early in the year Hasdrubal left Gades with his 30,000 and marched into the Baetis valley to recruit allies amongst the Turdetani, whose chief was Attenes, and made camp at Ilipa, near Seville. Hearing of these movements Scipio marched southwards, paused at Castula to gather reinforcements and, continuing towards Seville, made camp in full view of the Carthaginians. As the Romans set about fortifying this position they were attacked by the enemy cavalry under Masinissa and Mago and a fierce struggle ensued, until Scipio called in all the working parties and a strong column of armed men could be sent out to join the fighting. These skirmishes continued for the next few days until Hasdrubal had fully tested Scipio's strength, then the Carthaginian army marched out to do battle. The Romans followed suit but neither side made the first move to engage and after all day at arms the two armies withdrew into their respective camps. This happened on several successive days without a single missile being thrown, and each time the armies drew up in the same formations, the Romans facing the Carthaginians and Africans in the centre, the Spanish allies of both sides facing each other on the wings.

Scipio had learnt in the past that it was unwise to rely too heavily on new allies or mercenaries, yet if he discounted his Spanish allies he was outnumbered two to one. But the repetitive lining up for battle had sparked off an idea – he would switch the positions of wings and centre, thus preventing the Spaniards from being confronted by their fellow countrymen, and at the same time allowing him to use his strongest troops against the enemy's weakest ones. The evening before the battle Scipio ordered his army to rise before dawn, breakfast and be ready to leave camp by first light. The cavalry were to keep their horses bridled, with the saddle cloths on.

At dawn the next day all the Roman cavalry and light troops were despatched to drive in the Carthaginian outposts and harass the camp, while the legions and the Spanish allies formed up in their new positions on the open ground. There was considerable confusion at the Carthaginian camp during this time and Hasdrubal, ordering his cavalry and light troops out to engage their Roman counterparts, hurriedly led out his heavy infantry to meet the Roman main force, which was now marching forward in line of battle. Hasdrubal had no time for a prolonged scrutiny of the Roman line, which was probably mostly obscured by the light troops skirmishing, and, the numbers of the Spaniards and Romans being nearly equal, it is doubtful if a cursory glance at this time could have revealed the startling change in order of battle. For his own part, the enemy were fast approaching, apparently intent on doing battle at the earliest opportunity and, even had he wished to do so, there was no time to order a new line of battle. Each part of the Carthaginian army ran to take up the station it had occupied over the last few days.

The skirmishing of the cavalry and light troops continued for some time until, when the two main lines of battle were no more than 500 yards apart, Scipio ordered the recall sounded, opened his ranks, and allowed the cavalry and light troops to pass through to the rear. Here they were divided into two and posted in reserve behind the legions on each wing. Scipio must have waited for a considerable time before launching his main attack, for the wings did not meet in battle until noon. His men were rested, fed and fresh; the enemy had been roused abruptly from sleep, rushed to arms without a chance of food or drink and now stood to arms under the rising sun. Livy describes them by noon as being so weary of standing in the hot sun that many were leaning on their shields. Hasdrubal, of course, now knew the details of Scipio's changes, but the Carthaginian line overlapped the shorter Roman one, and Hasdrubal had more cavalry, plus the elephants, on the wings. If the change of positions did worry him, there was nothing he could do about it now – the armies were only 500 yards apart. Then Scipio began his great double oblique manoeuvre and envelopment. One begins to feel almost sorry for Hasdrubal, the greatest Carthaginian general after Hannibal and the other Hasdrubal, who may quite possibly have been literally mesmerised by what followed.

The complicated manoeuvres carried out by the Roman army may be split into three main parts for convenience, and these are best illustrated by the map.

(1) Each wing, including the cavalry and light troops, turned away from the centre and marched outwards to extend the Roman line so that it would overlap the longer Carthaginian line. The right wing was under Scipio, the left under Silanus.

(2) The wings then made a turn towards the enemy and advanced towards them. The various arms would have formed cohorts to perform this so that a front was always presented to the enemy. (Livy mentions three cohorts leading, supported by three troops of cavalry and light troops, the rest following in support.)

(3) The columns, nearing the enemy line, swung to left and right to form a new line outflanking the Carthaginian wings, infantry wheeling inwards, light troops and cavalry outwards, and at the same time the cavalry and light troops charged.

While the wings had been carrying out these manoeuvres the Spanish allies in the centre had advanced slowly, remaining always well behind the wings, and were still out of missile range when the action on the wings began. In the meantime, the Carthaginian army had remained stationary, no doubt watching this manoeuvre with wonder and puzzlement. Why the Carthaginian cavalry did not attack, or the wings waiting in line charge the Roman columns, can only be explained by

(1) they simply did not understand what was about to happen;

(2) the advance guard of three cohorts at the head of each of the columns, and the cavalry in close support, acted as a deterrent.

The Roman cavalry and light troops, hurling their missiles as they charged, stampeded the elephants in the enemy's front line and these threw into confusion the Carthaginian cavalry stationed behind them. Before the latter could rally again the Roman cavalry was upon them and in a short time had driven them from the field. Meanwhile the legions had made their assault on the Spaniards and the light troops now turned to attack the exposed flanks. The raw recruits from the Balearic Islands and the Turdetani, their stomachs empty and throats dry, found themselves fighting the cream of Scipio's army, each wing threatened with isolation from the rest of the Carthaginian army. Attacked on two sides, they began to give ground.

By refusing his centre, for a long time Scipio tied up the enemy's best troops, who were unable to go to the assistance of their allies for fear of the Roman centre charging them, yet were unable to charge the inferior Roman centre in return because of the victorious Roman cavalry which could have swept down on their thus exposed flanks. Eventually the centres did meet, but by now Hasdrubal's wings were under very heavy pressure and, tired and discouraged, the whole army began slowly to fall back towards their camp. The legions, seeing the enemy weakening, doubled their efforts and the movement to the rear became a retreat, then a rout as the Carthaginian army broke and ran for the security of its camp. The Romans were close behind them and might have stormed the ramparts, but at this moment a heavy rainstorm developed, the torrent of rain so heavy the Romans had difficulty in struggling back to their own camp.

The Carthaginians began building up the defences of their camp but during the night many of the allies, led by Attenes, began to desert, and rather than allow this to continue, Hasdrubal pulled out. The next day Scipio's cavalry found him and, inflicting heavy casualties on the Carthaginians, delayed the retreat so much that the legions were able to catch up with them. In the meantime many more of the Spaniards dispersed or defected. With the arrival of the legions, fighting broke out once more, the Carthaginians again losing heavily, and eventually only 6,000 survivors managed to break away and take refuge on an impregnable height in the nearby hills – without adequate water or indeed food supply. Hasdrubal fled to Gades, where ships awaited him, and, when more of the 6,000 deserted, was soon joined by Mago. The remainder of the 6,000 either slipped away into the surrounding countryside or defected to the Romans.

CONCLUSIONS. Scipio's manoeuvre had relied solely on the training and discipline he instilled in his troops. Abel to break his army into separate, flexible divisions in the face of the enemy, he had achieved two vital factors – surprise and concentration of forces. Ilipa was complete justification for his military reforms. Many of Carthage's most powerful Spanish allies now went over to the Romans and by the autumn the whole country was under Roman control. The battle had changed the whole nature of the war: from now on the battles would be fought in the Carthaginian homeland.

CASUALTIES. None of the sources give casualty figures for either side. We know only that 6,000 of the Carthaginians dispersed or defected at the very end. This means 48,500 were accounted for at Ilipa and during the retreat. Losses at Ilipa must have been heaviest amongst the Spanish allies, the remainder of whom deserted soon after. If we say perhaps 10,000 Spanish deserted, this still leaves casualties of 38,500 for Ilipa and the retreat. However, the important factor is that shortly after the battle an army of 54,500 men had ceased to exist.

BATTLE OF ILIPA

BATTLE OF ILIPA

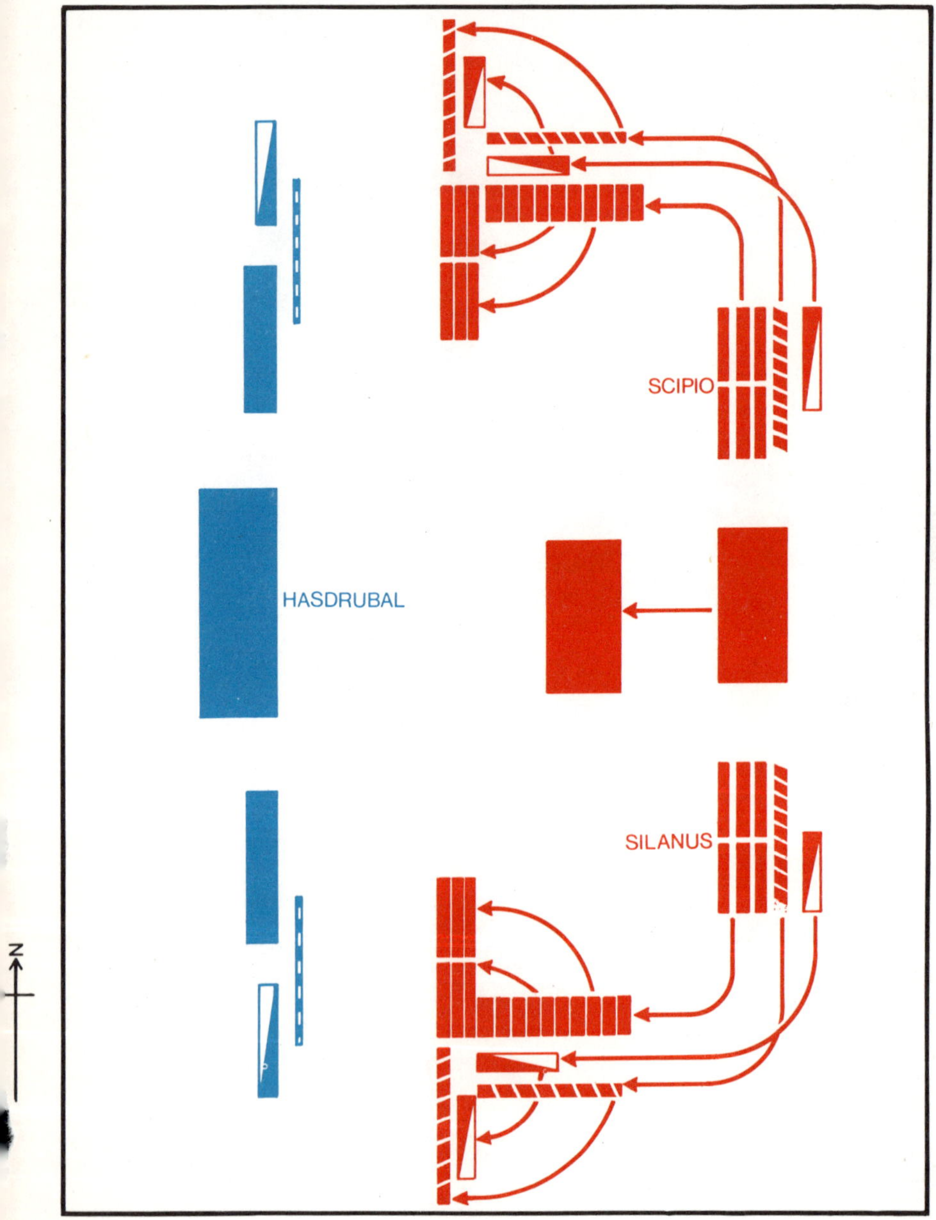

ILIPA N **1 man = 200, 1 elephant = 5**

Seen from the rear of the Carthaginian position, the Roman wings are just wheeling into their final positions. On each extreme flank are the Roman cavalry and skirmishers, just charging the Carthaginian elephants and cavalry. In the centre the Romans' Spanish allies lurk just beyond missile range and so tie up the mass of veteran infantry in the Carthaginian centre.

BATTLE OF ZAMA

Africa (Tunisia), October 202 B.C.

In 203 Carthage had begun peace negotiations and recalled Hannibal to Africa. The next year war broke out again but Scipio, outnumbered, hesitated to take the offensive until he could be reinforced by Masinissa, who was engaged in regaining his kingdom. Hannibal also hesitated to move from his base at Hadrumetum for, although joined by Mago's mercenaries and new African and Carthaginian levies, he was without cavalry and needed time to train this army of many nationalities. Towards the end of summer Scipio could wait no longer and set out to meet Masinissa, moving up the rich Bagradas valley, burning towns and corps to provoke Hannibal and deprive Carthage of an important source of food. Hannibal now had 2,000 cavalry but was waiting for more. At first he made no move but, under pressure from his Government, he finally marched to Zama to cut Scipio's communications, then reconnoitred from there, seeking to prevent Scipio and Masinissa from joining forces.

ROMAN FORCE. About 25,000 Romans and Italians, consisting of the V and VI Legions (disgraced at Cannae and banished to Sicily, where they were reorganised by Scipio), two legions of Italian allies; another legion of Etruscan volunteers; 6,000 Numidians with Masinissa. Cavalry: 2,000 Roman and Italian, 4,000 Numidian. Total : 37,000.

CARTHAGINIAN FORCE. 10,000 Gauls, Moors (from modern Morocco) and Ligurians, the latter being the most numerous: 2,000 Balearic slingers: 12,000 newly raised Carthaginian and African levies: about 12,000 of Hannibal's veterans from Italy, predominantly Bruttians: 4,000 Macedonians under Sopater. Cavalry: 2,000 Numidians under Tychaeus, perhaps another 2,000 from other Numidian tribes. 80 elephants. Total : 44,000.

TERRAIN. Scipio took up a strong position, where he could await the arrival of Masinissa, on a well-watered hill dominating the junction of the roads from Zama and Naraggara. Hannibal had no choice but to place his camp some three and a half miles away on a waterless hill. Apart from this strategic manoeuvering, the ground played no part in the battle, being fought on the bare and level plain between the camps where Scipio could make full use of his superiority in cavalry.

Before the battle could be fought Masinissa arrived with his 10,000 Numidians and Hannibal made a bid at peace. Scipio rejected the terms suggested and Hannibal was left with no alternative but to fight. His army was inferior in quality and training, and short of cavalry. He tried to counter these weaknesses by his dispositions. He could rely on Mago's mercenaries and on his own veterans from Italy, but not the African and Carthaginian recruits. Therefore, he sandwiched the latter between the other two, in three distinct lines, each line having orders not to allow those in front to pass through their ranks in any event. The Bruttians were about 300 yards to the rear – fulfilling a special role. Hannibal knew he could not hold his wings, so probably ordered his cavalry to 'run' from the battle on the basis that this way they could take the Roman cavalry off the field with them and that it would take much longer for the Romans to effect a decision in a running fight. He hoped this would give him time to defeat the legions, and to speed up this process he relied heavily on his exceptional number of elephants. The veterans of the third line were there not only as a reserve, but to guard his rear and to give the others, especially the new levies, a feeling of security, knowing their rear was guarded by an experienced body of men.

Scipio also made dispositions to deal with the special factors of the battle – those 80 elephants. Instead of the usual quincunx or chequer board arrangement, the maniples were arranged so that the gaps between them ran right through the three Roman lines, leaving alleys up which the elephants could run. The velites, after the skirmishing, had orders to retreat into these alleys and, if unable to withstand the elephants, to fall back to the gaps between the three lines. From here they could hurl their missiles at the passing elephants, then close up behind them after they had passed. Some of Masinissa's 6,000 infantry were also stationed at the rear of these alleys to deal with any of the elephants getting through, the remainder being mixed with the velites.

The battle opened with the usual skirmishing between the opposing light troops and cavalry wings, then Hannibal launched his elephant charge. The Roman velites fell back to the cover of the legions, suffering heavy casualties. The elephants were then met with volleys of javelins from the hastati, and by cheers and loud blasts on horns and trumpets. This caused many of the elephants to turn in panic and create a great deal of confusion on the flanks as they tried to escape. Seizing this opportunity, the Numidian and Italian cavalry charged home, sweeping the Carthaginian cavalry off the field of battle in a fast, running mêlée. In the centre many of the elephants ran into the alleys, where they were caught in a crossfire from the velites and, pushing on towards the rear, found themselves fired on from front and rear as well, and were soon killed off. Hannibal's elephant charge had failed, his cavalry were out of the battle, and the Roman line was unbroken. The lines of heavy infantry now began to close for the real contest.

For a while the Carthaginian mercenaries more than held their own against the legions but then superior equipment and training began to tell, the legionaries pressed forward steadily with shoulder and shield boss, and the mercenaries began to give ground. The Ligurians and Gauls bravely made repeated charges to push the line back but without avail and, seeing the success of the first line, those Romans to the rear began to push also and the increased pressure finally broke the stubborn resistance of the mercenaries. Most fled towards the flanks as instructed. but some, either in the heat of the moment or because they felt they should have been supported by the second line, turned and tried to force their way through the line of Carthaginians and Africans. The hastati had followed up and for a while there was a confused situation with Ligurians and Gauls fighting both sides, then the Carthaginian second line pressed forward and began to drive back the hastati. Scipio ordered up the principes and these managed to halt the Carthaginian advance, then push it back. Finally the levies broke and made for the flanks, pursued by the hastati, who broke ranks to do so. Scipio, seeing the principes following suit, had the recall sounded.

The ground between the armies was now so choked with dead and dying that it was almost impossible to advance in formation. Both sides therefore paused to clear their fronts of the wounded and to reorganise their lines. Hannibal needed some time to organise the survivors of his first two lines on the flanks of the Bruttians but Scipio, waiting for the return of his cavalry, was willing to allow him as long as he needed. Time was on the Romans' side. Seeing that the new Carthaginian line now overlapped and outnumbered his, however, Scipio had the hastati form a line in the centre and brought the principes and triarii up on each flank. This gave him a long line with maximum missile effectiveness when the two lines closed again, but there was the risk that the Carthaginians might punch holes in the line, which lacked depth. The struggle was then renewed and continued for some time without advantage to either side, and it is even possible that Hannibal might have won the day, but at long last the cavalry under Laetius and Masinissa arrived back on the field. Hannibal's Bruttians were now fully engaged at front and flanks, there was no opportunity to face up to the cavalry. The Romans and Numidians swept viciously into the rear of the Carthaginian line and the fiercely contested battle was transformed into a massacre. It was Cannae in reverse. Most of Hannibal's veterans fought to the death, the remainder were cut to pieces by the cavalry as they tried to flee across the open, flat countryside. Hannibal himself escaped with only a handful of horsemen.

CONCLUSIONS. Hannibal's defeat at Zama marked the end of sixteen years of war. Carthage no longer had an army and accepted Scipio's terms rather than endure a seige. Major terms were

(1) The destruction of 500 Carthaginian ships.

(2) A fine of £3 million, payable over 50 years.

(3) Reinstatement of Masinissa in his father's kingdom.

(4) No war to be fought outside Africa, and none within Africa without the permission of Rome.

The loss of the ships, and the imposition of a large Numidian kingdom on their borders, spelt the end for Carthage. For them it was the end of an empire: for the Romans the beginning of one.

CASUALTIES. Roman : Most sources quote 1,500 but the figure given by Appian, 2,500 is considered nearer the mark. This does not include Masinissa's losses, which were more numerous, so the total Roman losses could have been as high as 6,000. **Carthaginian :** 20,000 killed and almost as many taken prisoner – a total of nearly 40,000 out of the 44,000 engaged.

BATTLE OF ZAMA

BATTLE OF ZAMA

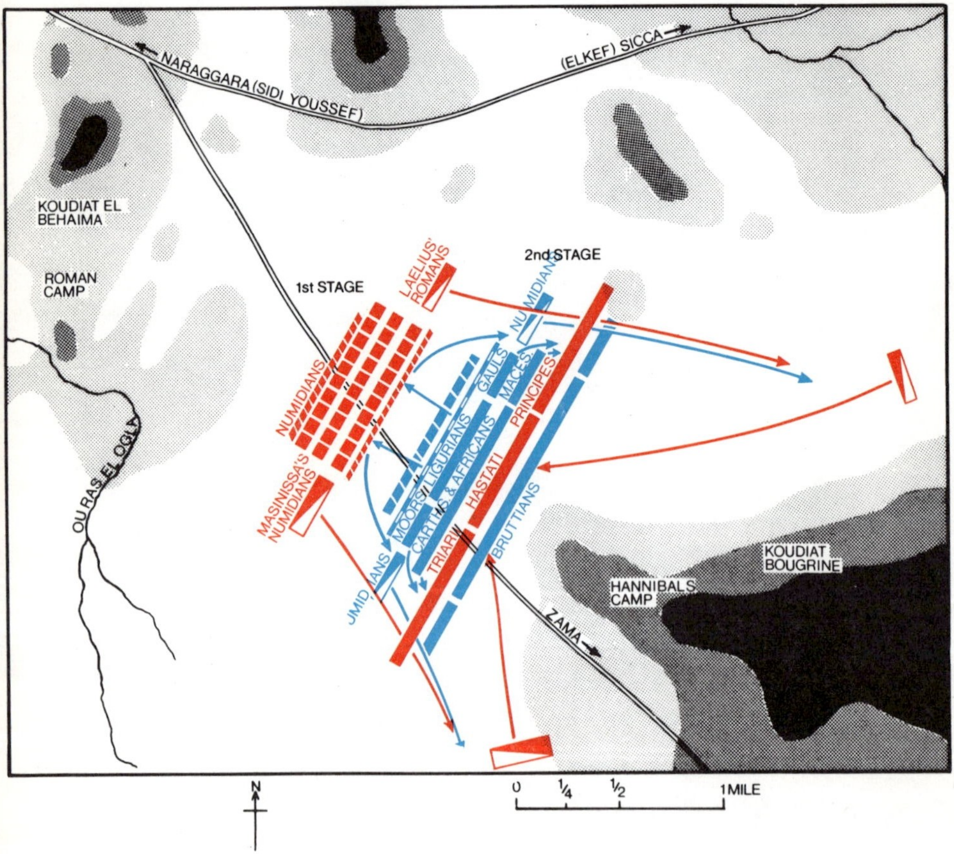

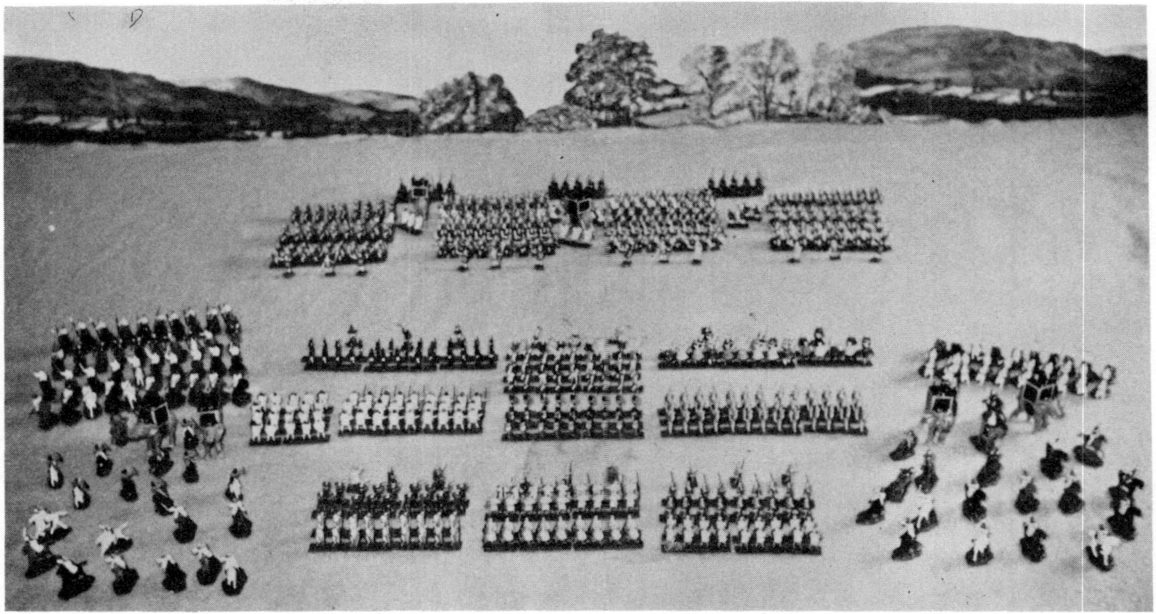

ZAMA N **1 man = 100, 1 elephant = 10**

Taken from the rear of the Carthaginian forces, this photograph shows the battle of Zama at the time when the elephant charge has just failed, the Carthaginian cavalry are being driven off in confusion, and the main force of infantry are beginning to advance to do battle. The three lines of the Carthaginian force are very plain here. Remnants of the velites and Numidian infantry can be seen dealing with the remaining elephants between the cohorts. Note especially the heavy mass of cavalry on the left. There were two to one odds against the Carthaginians and they had no chance here: Hannibal's orders to his cavalry made the best of a bad situation.

ORGANISATION GUIDE

Roman Legion c. 216 B.C.
(5,280 men, nominal 6,000)

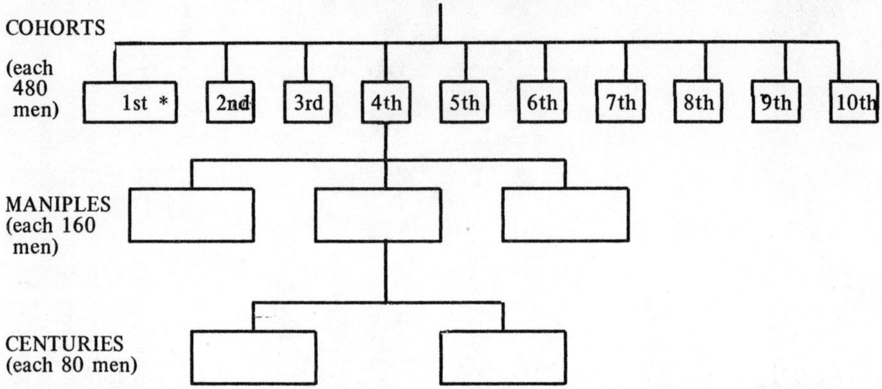

* The 1st Cohort, which included H.Q. personnel, was of double strength.

Cavalry. Each legion was accompanied by 300 Roman cavalry, divided into 10 turmae or squadrons, each commanded by 3 decuriones. Allied cavalry, theoretically totalling 900, was divided into 3 alae.

LINE OF BATTLE

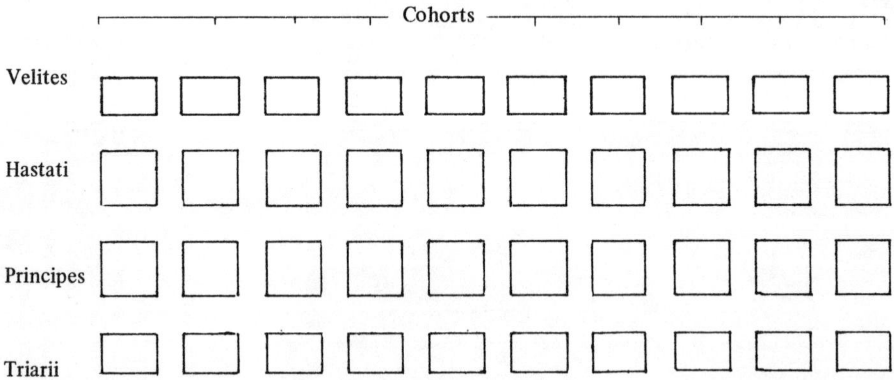

EQUIPMENT

Velites: Round shield, 3′ diameter; sword*; 2 4′ javelins with 9″ metal head; helmet without crest.

Hastati: Scutum 4′ x 2½′; sword; 2 pila, thrown in volleys before coming to close quarters with enemy; bronze helmet with 18″ plume; small bronze breast-plate, wealthier citizens may have worn mail armour; greaves.

Principes: As hastati but with a long spear or pike (hastae) instead of the pila.

Triarii: As principes.

Cavalry: Shield, lance, sword, helmet, greaves and lorica or corslet.

* At the time of the 2nd Punic War the Romans used a short thrusting sword. At Cannae they met the long unpointed, cutting swords of the Gauls and in Spain the Iberian sword used for both stabbing and cutting. After the fall of New Carthage, the Legions under Scipio adopted the gladius hispaniensis (2′ long and 2″ wide) a combination of the cutting and thrusting properties.

OFFICERS

Each century had 1 centurion, 1 optio and 1 vexillarii (standard bearer). H.Q. was made up of a consul or legatus legionis, six military tribunes, aquilifer (eagle bearer), and the equivalent of the modern service and supply units – medical, engineering, clerical, orderlies, armourers etc.

All figures used for the reconstruction are Airfix, with many conversions:-

Romans – Roman and Italian cavalry and infantry.

Ancient Britons – Gauls, Ligurians, Carthaginian and Bruttian infantry, Balearic slingers.

Indians – African, Libyan and Moorish infantry, Spanish and Gallic cavalry, Roman auxiliary cavalry

Arabs – African heavy infantry, Numidian infantry and cavalry.

Robin Hood – Gauls, Libyans, Spanish and Balearic islanders.

War elephants were Britain's baby elephant with a howdah added.

